Snag It!

Who Needs a Dead Tree?

by Mary Holland

Have you ever heard of a snag?

A standing tree that is dead or dying is called a snag. It can be standing in water or on land.

When a tree dies, its leaves and bark fall off.

A snag is a very important part of the forest on which many animals depend.

All kinds of animals use snags. They build nests and raise young in them. They rest (perch) in them. Animals store food in snags. They shelter from snow, wind, and rain inside snags. Some animals even use snags to hide from animals that might eat them (predators).

This green heron perches on a snag looking for a fish or a frog to eat.

Hawks, herons, eagles, ospreys, and owls use tall snags as hunting perches. They can spot mice, voles, fish, and other prey from high above the ground.

Bald eagles have excellent eyesight and can see prey well over a mile away.

Insects, mosses, lichens, and fungi can be found in and on snags. They provide a wide variety of food for animals.

Many insects lay their eggs in dead trees. When the insect eggs hatch, the young insects (larvae) live inside the tree until they grow wings and fly away.

Insect-eating birds, like this red-breasted nuthatch, drill into snags with their strong, pointed bills to eat the larvae.

Woodpeckers nest inside living trees and snags. They drill holes (cavities) with their strong bills. They raise their young inside these holes where the young are protected from the weather.

The young pileated woodpecker looking out of its nest hole is about to fly out of its nest and leave home (fledge).

Birds aren't the only animals that nest in snags. When a branch falls off a tree, it often leaves a large hole. Raccoons use these holes as nests. Sometimes the holes are big enough for a family of six or more raccoons.

Porcupines den in snags too. They also den in rock piles or in living trees. Porcupines mostly use dens in the winter for protection from snow and wind.

Animals often seek shelter inside hollow snags.

Flying squirrels are active at night (nocturnal) so we don't see them very often. They glide from tree to tree looking for seeds, nuts, fruit, and fungi to eat. They spend their days sleeping inside snags.

Like flying squirrels, bats may also seek shelter in snags during the day. They sleep (roost) behind loose bark or in the snag's hollow center where they are protected from weather and predators.

Some bats hibernate inside snags in the winter.

As a tree rots, the inside turns into sawdust. Mice, voles, snails, slugs, spiders, beetles, and many other creatures live in this rotting wood.

Snakes, like this eastern gartersnake, search for food inside snags. They also often lay their eggs in rotting snags. The sawdust keeps the eggs warm.

The insides of snags are shady and moist. This is a good place for insects and worms to live.

Salamanders, like this eastern red-backed salamander, can often be found inside snags looking for insects to eat.

Snag branches are often sunny because there are no leaves for shade. These branches are a perfect place for birds to sit and rest.

These turkey vultures are perched high on a snag. They spread their wings to dry their feathers and warm their bodies.

Foresters often call snags "wildlife trees." Sometimes snags have more wildlife, like this screech owl, living in them than live trees!

Think of snags like apartment buildings. Salamanders, slugs, spiders, beetles, snakes and different kinds of birds and mammals can all find food and raise their families in one snag.

Because snags are so important to so many animals, they should not be cut down. Even though they aren't alive, snags provide a home, food, and safety for many animals.

For Creative Minds

Pileated Woodpecker Food and Nest Holes

Pileated woodpeckers drill holes in snags to find ants, beetles, and other insects to eat. These holes are usually longer than they are wide.

They also drill holes to make nest cavities. These holes are usually round.

Can you find one of each kind of hole in the snag on the left?

Birds & Snags: Did You Know?

Birds that eat insects, like woodpeckers and nuthatches, depend heavily on snags as a source of food. These birds help to control unwanted insect pests.

Over 85 species of North American birds use cavities in dead or deteriorating trees.

All woodpeckers lay white eggs. Because they are hidden inside the tree, the eggs don't have to be camouflaged.

Birds that nest in cavities tend to nest earlier in the spring than other birds. Their eggs are well protected from the cold, snow and rain.

There are two kinds of cavity-nesting birds, primary and secondary. Primary cavity nesters, including all woodpeckers, excavate their own cavities. Secondary cavity nesters do not excavate their own cavities–they use those made by primary cavity nesters.

Which Animals Use Snags?

Dead trees provide habitat for more than 1,000 species of wildlife in the United States. Which of these animals find food, shelter, or a resting place in snags?

beetle larva

long-tailed weasel

salamander

red-tailed hawk

honeybees

spider

black bears

barred owls

wood ducks

Answers: all

How Might These Animals Use Snags?

Animals may use snags for several reasons. How do you think these animals use snags?

nest	roost (sleep)	perch (rest)
hibernate	hide	find food

screech owl

bald-faced hornet

double-crested cormorant

eastern chipmunk

mourning cloak butterfly

spider

Possible Answers: Screech owl-roost, nest, perch; Bald-faced hornet-hibernate; Double-crested cormorant-perch, roost; Eastern chipmunk-hide, find food; Mourning cloak butterfly-hibernate; Spider-hibernate, find food

Nests in Snags

Many birds nest in snags, including swallows, herons, owls, and even ducks!

Deep inside this snag a female wood duck has laid her eggs. She sits on them to keep them warm until they hatch.

The young ducklings must jump out of the snag (fledge) when they are old enough to join their mother.

If the snag is standing in water the ducklings land in the water when they jump.

If the snag is in the woods the ducklings bounce like tennis balls on the forest floor when they land and then follow their mother for up to a mile to find water.

True or False?

1. Trees can provide more habitats for wildlife dead than when they are alive.

2. In total, more than 100 species of birds, mammals, reptiles, and amphibians need snags for nesting, roosting, shelter, denning, and feeding.

3. Hollow snags are very valuable in winter as they are used by many species like squirrels, raccoons, owls, and bears for denning and roosting.

4. Many snags are formed when trees break during strong windstorms.

Answers: all are true

To all the landowners who strive to enhance wildlife habitat by preserving snags on their property.— MH

Library of Congress Cataloging-in-Publication Data

Names: Holland, Mary, 1946- author.
Title: Snag it : who needs a dead tree? / by Mary Holland.
Description: Mt. Pleasant, SC : Arbordale Publishing, LLC, [2025] |
  Includes bibliographical references.
Identifiers: LCCN 2024033717 (print) | LCCN 2024033718 (ebook) | ISBN
  9781638173267 (trade paperback) | ISBN 9781638173304 (ebook) | ISBN
  9781638173342 (epub) | ISBN 9781638173380 (pdf)
Subjects: LCSH: Snags (Forestry)--Juvenile literature. | Forest
  ecology--Juvenile literature. | Forest animals--Juvenile literature. |
  CYAC: Dead trees. | Forest animals. | Forest ecology. | LCGFT: Picture
  books.
Classification: LCC QH541.5.F6 H65 2025 (print) | LCC QH541.5.F6 (ebook)
  | DDC 577.3--dc23/eng/20240802
LC record available at https://lccn.loc.gov/2024033717
LC ebook record available at https://lccn.loc.gov/2024033718

This title is also available in Spanish: ¡Troncón! ¿Quién necesita un árbol marchito?
Spanish paperback ISBN: 9781638173441
Spanish ePub ISBN: 9781638173564
Spanish PDF ebook ISBN: 9781638173625
A dual-language read-along is available online at www.fathomreads.com.

English Lexile® Level: 650L

Bibliography
Holland, Mary. Naturally Curious: A Photographic Field Guide and Month-by-Month Journey through the Fields,
  Woods, and Marshes of New England. Second Edition. Trafalgar Square Books. North Pomfret, VT, 2019.
  Winner, National Outdoor Book Award.

Printed in the US
This product conforms to CPSIA 2008

Arbordale Publishing, LLC
Mt. Pleasant, SC 29464
www.ArbordalePublishing.com